ABOVE : *One of a pair of steam ploughing engines.*
LEFT : *An elaborate steam powered cultivator made in 1857 by Crosskill. Its weight was too great for it to be widely used.*

VINTAGE FARM MACHINES

John Vince

Shire Publications Ltd.

CONTENTS

ABOVE: *A horse hoe with swing steerage; one of the items from 'Mary Wedlake's Priced List of Modern Farming Implements' 1850.* OPPOSITE: *A single-handed Norfolk plough, about 1880. (University of Reading Museum of English Rural Life.)*

INTRODUCTION

Primitive farmers had to depend upon their own strength to prepare the soil and gather the harvest. An unknown genius invented the plough and arranged things so that oxen and horses could be made to bear a large part of man's burden. For more than a thousand years things remained unchanged and practically every other agricultural task demanded hand labour.

In the eighteenth century techniques began to improve and, encouraged by economic influences like the enclosure of the common fields, farmers began to explore ways of saving man-power. Experiments with agricultural machines were indirectly affected by the advances made in the textile industry. The use of cast iron to make gear wheels was of considerable importance and had a profound influence on technical innovation.

Gradually, machines became involved in sowing seed, haywork, harvest and even threshing. In time there were very few jobs about the farm that a machine could not undertake. Machinery received a new impetus when steam power became available and partly replaced the horse. Some machines that were introduced in the 1840s hardly changed in appearance during the next 100 years, so sound were the first designs of those Victorian agricultural engineers.

In this present age of advanced technology one must not overlook the historical importance of the farmer's older machines, for there are a good many veteran binders, of some sixty summers, still in use!

Magnificent examples of Victorian and Edwardian engineering remain gathering cobwebs in half-forgotten barns. When they do appear briefly at farm sales the scrap merchant is often the only bidder. The following pages provide a short pictorial guide to some of these technical masterpieces.

ABOVE: *A Fowler 14 nhp 12 tons single cylinder plough engine working with a cultivator in Northamptonshire in 1915.*
OPPOSITE: *Ploughing with oxen, about 1880 (University of Reading Museum of English Rural Life).*
BELOW: *A Wallis and Steevens 1908 single cylinder light steam tractor ploughing in 1909. The engine is moving backwards and the plough at the rear is raised clear of the ground (Wallis and Steevens).*

POWER

As long ago as the 1840s steam engines were beginning to be used on the farm. These early portable engines had to be pulled from place to place by horses. At that time boilers were worked at low pressures and therefore required large cylinders to develop the required power. In the 1850s the development of high pressure boilers reduced the size and weight of steam engines. When engines, which could propel themselves, came into use steam power became an alternative to the horse. The horseless plough must have seemed as unworldly as the moon to Victorian countrymen. But it did not take long for the new ideas to gain acceptance in farming circles.

Only rich farmers could afford the considerable expense of a set of two ploughing engines and all the ancillary equipment (such a set would cost about £1,800) and even then a proprietor had to be sure that his machine did not stand idle. To obtain an adequate return on the capital invested, ploughing sets were sent out to work on contract to many other farms. With the introduction of the steam plough a new kind of journeyman came into being and writers like Flora Thompson in *Lark Rise to Candleford* have noted the details of his attitudes and way of life. For almost seventy years (1850-1917) the steam plough co-existed with the horse which it never really replaced, especially on the smaller farms.

Steam power also brought about a revolution in the design of farm machinery. Metal ploughs of immense proportions partly replaced those fashioned in wood and iron by village craftsmen. The monsters of the steam age endured until they were laid low by the internal combustion engine.

The internal combustion engine was introduced to agriculture before 1910. It did not take long for the new and lighter tractor —able to manoeuvre over the ground it cultivated—to present a serious challenge to the cumbersome steam plough restricted by its weight to field edges. Vintage tractors are now recognised as an important part of agricultural history.

TOP LEFT: *Allis Chalmers Model B tractor manufactured in 1935.*

CENTRE LEFT: *An International 8-16 tractor of 1919 vintage.*

BOTTOM LEFT: *A Bamford Market Gardener tractor of 1920.*

ABOVE: *A ploughing match in progress in Berkshire during the late 1960s.*

RIGHT: *A John Deer A.R. (2 cylinder) tractor of 1936.*

TOP LEFT : *A Sussex breast plough.*

BELOW : *A wooden skim plough, the horse-drawn alternative to the breast plough. This one was sold at a farm sale in 1971 for £9.*

TOP RIGHT : *Details of breast plough head. This example has its right hand edge raised to make a vertical cut through the turf. In some areas the left hand edge was preferred.*

PARING AND BURNING

Before chemicals were used to destroy weeds or insects the farm worker had to do things the hard way. The old method was by paring, which was to remove a thin slice of the topsoil in which weeds and larvae were to be found. The top layer was then burned and reduced to ash which was worked back into the ground. By 1880 the method was considered to be out of date but it did not disappear until well into this century. The tool (a breast plough) used for paring looked like a long spade. Various local names were used for it : a flautcher spade (Scotland), a cast-cutter (Kent), a velling spade (Cornwall). The illustration above shows it in use.

PLOUGHING

Although one of the oldest agricultural implements, the plough has changed very little over the years. To make a furrow a plough makes two cuts in the soil. A vertical cut is made by the coulter and a horizontal cut is made by the share. The forward movement of the plough allows the furrow slice to be turned sideways by the shaped mouldboard (the breast).

Until the all-iron plough was introduced in 1800 ploughs had been fashioned in wood; ironwork had been confined mostly to the share and the coulter. Wooden ploughs were still made, of course, and their use persisted even after the tractor had been invented.

The old wooden ploughs were made-to-measure affairs constructed to meet the particular requirements of local conditions. Once the business of ploughmaking became the concern of the factory the plough's features were subjected to standards imposed by the ironmasters. The age of spare parts had arrived and the new iron ploughs were constructed so that different components could be fixed or removed at will. This arrangement had advantages for the farmer. One basic plough could become several different tools if a selection of attachments was also purchased. In this way capital costs could be reduced.

Local preferences for particular details like the shape of the mouldboard could be met by providing a selection of components with different dimensions. Some manufacturers, like Ransomes of Ipswich, employed their

LEFT : *Wheel-plough with wooden beam and handles. The large wheel ran in the furrow. Both wheels have scrapers to keep them clean.*

BELOW : *Detail of the plough above showing the adjustable coulter (1), the winged share (2), the ground wrest (3), and the mouldboard (4).*

own teams of ploughmen. Their task was to visit ploughing matches up and down the country in order to discover the different regional needs and to advertise the superiority of the iron plough. Great importance was attached to the precise shape of the mouldboard as it could be made to throw the furrow over with a sudden or a slow movement. The difference was clear to see in the broken or unbroken furrows left in the plough's wake. Certain conditions required land to be broken up and left for the wind, rain, and frost to reduce. The shape of a mouldboard can tell us a lot about the work a particular plough could perform.

Many ploughs have a fixed mouldboard normally on the right hand side. A ploughman using a plough of this kind could turn the plough team round at the end of each furrow and return in the opposite direction. If he did this the second furrow would again fall to the right and leave a wide gap as the two furrow slices would be facing away from each other. One object of ploughing is to produce a surface that can readily be made flat by the harrow.

Open furrows were made at intervals to help drainage but the strips in between had to be as flat as possible. This requirement determined the way the ploughman set about his work. If each field was ploughed in a series of separate strips the first furrow could be made at its centre. The return furrow was cut on the right of the first and the two furrow slices leaned together like the two sides of a roof. For the remainder of the strip the ploughman worked up one side and then down the other. The result was a strip ploughed in the fashion shown in the illustration opposite.

There were various ways of ploughing which depended upon local tradition and geography. To overcome the disadvantages of a plough with a fixed mouldboard the turn-wrest plough was invented. Its mouldboard could be changed over at the end of a furrow so that the furrow made on the return journey fell on top of the previous slice.

Wooden turn-wrest ploughs were in use before 1800 but the makers of iron ploughs soon copied their features. Turn-wrest ploughs are also called one-way ploughs.

a   b   c

d   d

This name is sometimes misleading as one could expect the plough with a fixed mould-board to have been given such a name. The term, however, refers to the way the furrows fall. This is why iron one-way ploughs have two mouldboards!

Most agricultural writers agreed that the most suitable length for a furrow was be-tween 220 and 250 yards. The old sub-division of the mile—the furlong (furrow-long) was 220 yards. The time taken to turn horses at the end of a furrow added up in a significant way. In a field with 50 turns per acre turning time was calculated at 30 minutes. A hundred turns accounted for 1 hour 15 minutes and to this time had to be added for resting. Pulling the plough was hard work and some soils needed a plough team of four horses. During the working day the horse walked a consider-able distance even if the speed of the plough team was a slow 2 m.p.h. Each horse had to exert a pull of something like 280 lb. and it is easy to appreciate why the operation was carried out at a slow pace. To plough an acre involved walking from ten to twelve miles in the field. An acre a day was a good stint even if the field was not far from home.

The ploughs which survive today and

ABOVE : *Diagram to show cast, yoked or coupled ridge—a.b. furrow slices lying to the right; b.c. furrow slices lying to the left; d.d. open furrows between ridges.*

BELOW : *A wheel plough with a wooden beam in use at a rally at Marsh Gibbon, Buckinghamshire in 1972.*

appear at farm sales or quietly rot away in secluded hedgerows are seldom more than half a century old. Their characteristic features however appeared on the ploughs of our medieval ancestors. They provide an important link with those distant years when the English landscape was patterned with open common fields in which each villager had a share of the narrow strips to till himself.

The importance of the plough is also illustrated in the survival of various sur-names probably derived from it. Among these we can include: Rice or Rist (an alternative name for a mouldboard derived from turn-wrest); Mold (from mouldboard); Slade (the base of the plough—the chip—which slides along the furrow); Buck (the socket which carries the share); Eakes (the axle of a gallows — Norfolk or Kent — plough); Pratt (a U-shaped iron projecting from the eakes to which the draught chain was attached); and, of course, Plowright and Plowman.

BELOW : *A Sussex swing plough (locally known as a foot plough) with an iron 'foot' at the front end. The mouldboard is made of ash and called the 'wrist'.*

BOTTOM : *Detail of the swing plough showing (a) beam, (b) sheath, (c) chip,* (d) buck, (e) share—note the old style chisel edge, (f) stump, (g) coulter, (h) hog. An ash stick can be seen wedged between the upper part of the sheath, the coulter and the bracket (j) fixed to the beam. This stick was moved into the alternative position when the coulter was set on the other side of the buck.

TOP RIGHT : *A plough suitable for hard land. It could turn a furrow 14 inches wide and 9 inches deep.*

TOP CENTRE RIGHT : *This plough with its long mouldboard was made by E. H. Roberts of Deanshanger, Northants (G. Weatherhead Collection).*

LOWER CENTRE RIGHT : *Like the plough at the top of the page this example has a spreader projecting from the rear of the mouldboard. This device was designed to level off the soil which had been turned over by the breast.*

BOTTOM RIGHT : *A double-furrow plough. (Norton's Farm Museum, Sedlescombe, Sussex.)*

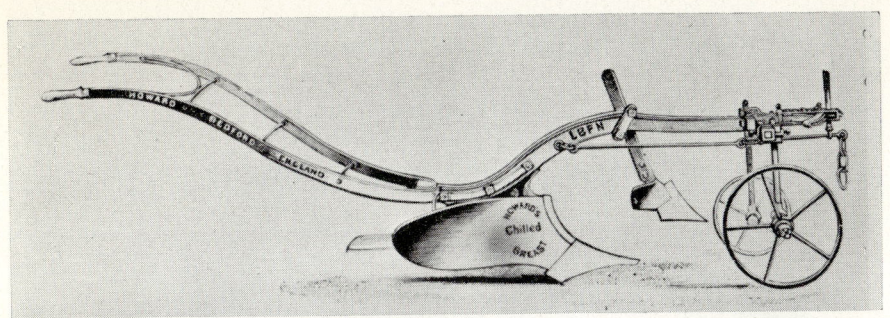

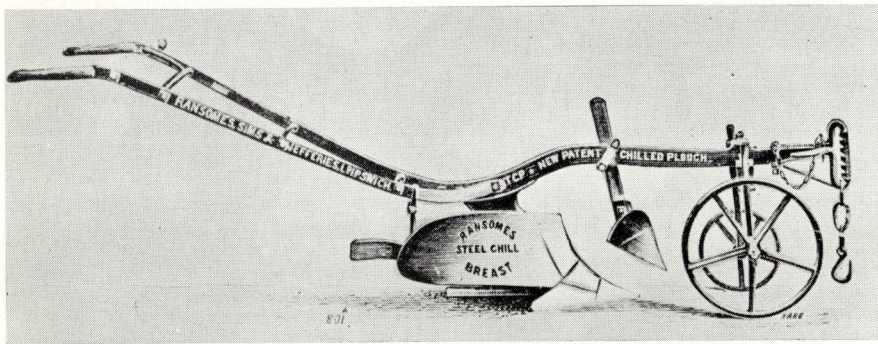

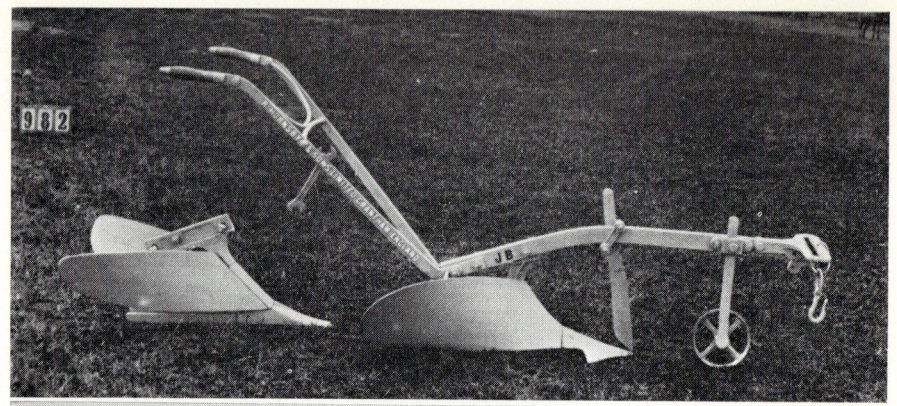

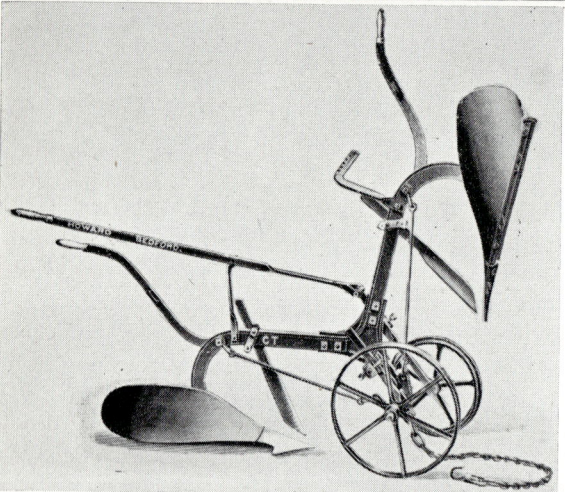

TOP : *A Hornsby plough showing the alternative ridging device. (Reading University Museum of English Rural Life).*

LEFT : *A balance turn-wrest plough made by Messrs. Howard of Bedford.*

BOTTOM : *A furrow press. The shafts have been removed and a tractor tow bar fitted.*

STEAM PLOUGHING

When steam power was applied to ploughing, experiments were made with single machines and with engines working in pairs. The latter system proved to be the most acceptable although it required more capital investment. There were disadvantages in the use of this heavy equipment. The power of the engines made it possible to plough much deeper than the horse plough. In some places inept use of the equipment did in fact spoil land which had only a thin layer of workable soil. Ploughing too deep caused the good layer to be buried under the subsoil with disastrous results to subsequent crops. A great deal depended upon the skill of the enginemen and their understanding of the land.

The double engine sets were very efficient when worked by skilled hands. An engine was positioned at each side of the field and the immense plough was pulled from one to the other in turn. In very large fields the second engine was often out of sight and the enginemen communicated with one another by whistle signals. A six furrow plough could work 12-14 acres a day. The horse plough could manage little more than one acre.

The plough was only a part of the equipment used. Other implements included an eleven-tined cultivator which could rip its way through as much as 25 acres a day. Harrows, too, could be worked very satisfactorily by plough engines. Another useful device was the mole-drainer. It usually had a triangular frame and at the end of its substantial coulter (approx. 12 in. by 2 in. in cross section) there was a 'mole' shaped like a bullet. As the machine moved forward the coulter was lowered to the appropriate depth and the 'mole' left a three-inch hole which allowed rainwater to drain away. Above the channel left by the 'mole' the coulter made a vertical slit into which surface water could find its way.

Steam ploughing added a new dimension to agriculture by introducing speed into a task that had never moved at more than 2 m.p.h. It showed that it was possible to make economies in manpower too and it began a trend which is an important part of the farm accountant's thinking today.

ABOVE : *A six-furrow anti-balance plough reaches the end of a bout. It will be noticed that a considerable headland remains unploughed. When the steam plough had finished it was not uncommon for the headlands to be ploughed by the horse! The anti-balance plough was invented by Fowlers in 1885. A plough which had its weight arranged symmetrically tended to ride up in use and not dig as deep as it should. This happened because the half that was up in the air was inclined to bring the implement back to its point of balance—i.e. with both sets of shares above the ground! To eliminate this tendency the Fowler anti-balance* plough had its undercarriage arranged so that the axle could move forward beyond the point of balance. The pull of the cable controlled the position of the axle. Once the axle had moved out of centre one half of the plough became heavier than the other and the tendency to ride up was overcome.*

OPPOSITE : *Two passengers help this anti-balance plough to bite harder. The trailing rope can be seen passing through the eye at the tail end of the plough frame.*

BELOW : *The winding drum could hold at least 440 yards of cable.*

STEAM DIGGING

The plough evolved from the humble spade and in the last century agricultural engineers became interested in steam machines designed to dig instead of plough. Harold Bonnett in 'Saga of the Steam Plough' records the names of several men who explored ideas for digging machines. A. Atkins of Chepstow invented a rotary digger hauled by a cable in 1843. James Usher of Edinburgh patented a mobile steam engine with a revolving digger mounted at the rear in 1849. Three years later John Bethell made a rotary digger. At Buckingham in 1857 a Thomas Rickett constructed a steam-powered rotary digger. In the same year J. Bethell of Westminster patented a wheeled digger. Another digger with hydraulic rams for raising and lowering the tines was invented by W. H. Nash.

In 1871 the Royal Agricultural Society of England held trials at Wolverhampton for steam cultivating machinery. The award of a medal at such an event was the ultimate accolade for the fortunate manufacturer. Prizes for steam machinery were sparse compared with the awards given for stock rearing. This may have had the effect of slowing down the pace of inventors' experiments. An important agricultural writer noted that 'It is a matter of regret that Agricultural Societies, especially the Royal of England, give so little encouragement to the makers of agricultural machinery.' Nine years elapsed before the next trials were held— at Carlisle in 1880. There the vast digger invented by Mr. Darby of Chelmsford made its debut and won a special prize. The merit of his digger was its fifteen per cent economy in the power needed to perform the same work as the plough tackle which won first prize at Wolverhampton in 1871. The Darby digger could cover about 10 acres a day.

In 1918 an article appeared in 'The Electrical Review' describing the Zimmerman Plough which employed electric motors instead of the steam engine. The idea does not seem to have progressed beyond the experimental stage and any chance of success it may have briefly enjoyed vanished with the development of the tractor.

ABOVE: *The digger shown here was de-signed by Mr. Frank Proctor of Stevenage, Herts. A traction engine was adapted to provide the necessary power for the crank-shaft on which the three forks were located. The traction engine was therefore a dual purpose machine as it could still be used for all kinds of other work. Mr. Proctor claimed that his digger could work 10 acres a day—the same as the Darby digger. Proc-tor's digger consumed about 11 cwt. of coal during a working day and required the attendance of two men. Its cost was £800 compared with the Mr. Darby's £1,200.*

OPPOSITE: *This combined cultivator, seed drill, roller and harrow was built by Fowlers of Leeds (c. 1907). It was probably one of the heaviest implements ever used on the farm. (University of Reading Museum of English Rural Life.)*

BELOW: *An early tractor pulling a double furrow plough fitted with disc coulters.*

ABOVE: *This water or liquid manure spreader is fitted with a pump identical in design to those used for domestic purposes. The tank is fixed centrally above the axle. This made it self-balancing when in motion and easier for the horse. Tanks of this type were made in a wide variety of sizes from 90 to 220 gallons. The largest size cost £20 in the 1890s. The pump was an optional extra at £4:5:0.*

LEFT: *This small crop sprayer probably dates from the 1920s. It was intended for use in the orchard.*

BELOW: *After ploughing came the harrowing to make the surface level for the seed drill. This ancient wooden example is interesting as its curved members were obviously all sawn from the same bough. (Norton's Farm Museum, Sedlescombe, Sussex.)*

SOWING THE SEED

We do not know how the prehistoric farmer set the precious seed he had so carefully saved from the previous harvest. The sower of Biblical days, however, walked across the ploughing and cast the grain with an unchanging rhythm to right and left by hand. One great disadvantage attached to broadcasting seed in this manner: some seed fell to the bottom of the furrows, the remainder was scattered on the crest or sides of each furrow slice. The irregular distribution meant that those seeds that fell into the bottom of the furrows grew in clumps, while others resting on the top grew in random isolation.

Sowing by mechanical means allowed the seed to be placed at an even depth and spaced at regular intervals in precise rows. Broadcasting, though cheaper, could never match the machine's precision.

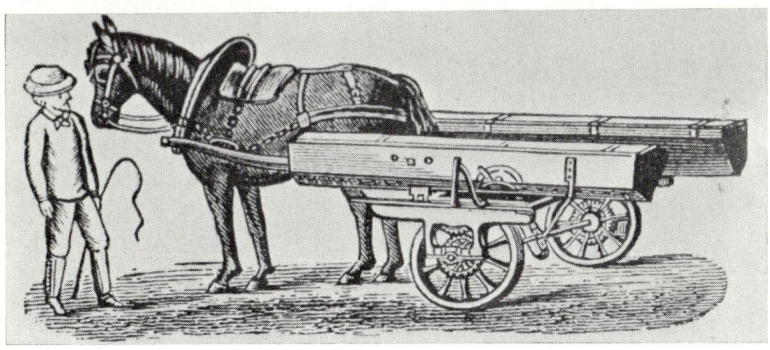

LEFT : *A later improvement was the 'fiddle' which must be one of the smallest farm machines. Its popular name comes from the action of the sower in moving the bow backwards and forwards. This broadcast sower was invented in North America and introduced into England by Mr. J. H. Newton of West Derby, Liverpool. The machine consists of a light wooden box which is carried under the left arm. A canvas bag for the seed is fixed to the top of the box. The seed wheel is attached below the box and round its spindle passes the thong which forms the bow string. As the operator pushes the bow to and fro the wheel rotates in alternate directions. The internal mechanism feeds the wheel with seed and the motion expels the grain for anything up to twelve feet on each side. With this simple device a sower could cover as much as four acres per hour.*

BELOW : *A horse-drawn version of a broadcast sower. In use the two seed chests were arranged in line with the axle. Gear wheels at each end of the axle provided the driving power for the internal mechanism.*

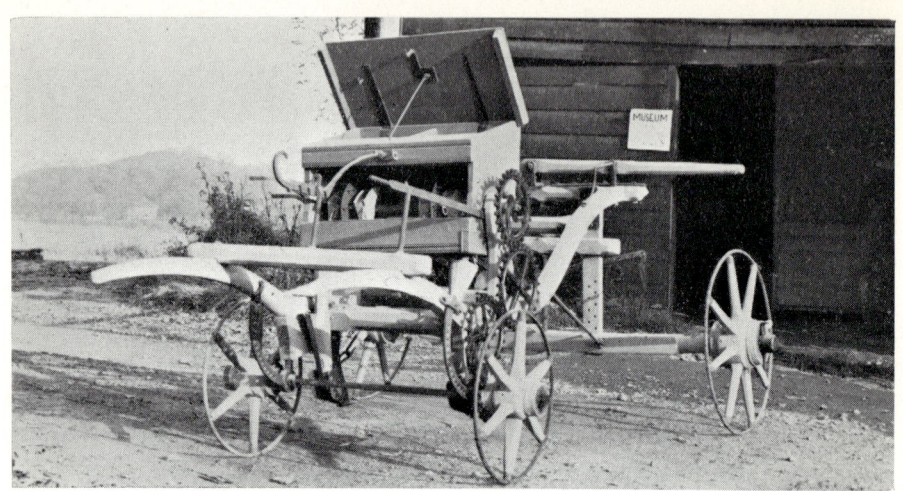

ABOVE: *A Kent Pattern 4-share Corn, Bean and Seed Drill made by Henry S. Tett & Co. Ltd. of Faversham, Kent, and now at Norton's Farm Museum, Sedlescombe, Sussex. The drill represents a very old design that could have been in use during the first half of the last century, but this particular example was probably made in the 1920s. The drill was steered from the rear (front of photograph) and the handles enabled the operator to lift the coulters clear on the headlands. The front (wide) axle was arranged to allow the wheels to be adjusted for sowing rows 7 ins, 9 ins, 10 ins, 11 ins, 18 ins, 20 ins and 22 ins apart. The rear axle provided the drive for the simple mechanism by means of the cast gear wheels which can be clearly seen. Each land wheel has spokes terminating in ferrules to which the square section hoop tyre is attached. The purpose of this kind of tyre was to leave a clear mark on the ground for the steersman to follow after each turn-round on the headland.*

Details of the Kent drill opposite. ABOVE: A view of the seed box which shows how the gears drive the cup feed mechanism. This very simple arrangement was highly effective and copied by many other makers. The seed was allowed to fall from the hopper above into the seed box. Then the rotating discs with their cups—rather like miniature spoons—lifted the seeds. When a cup reached the top of the circle the seed fell into the narrow hopper and began its journey earthwards. RIGHT: The undertins through which the seed passed on its downward way. The upper tin was called a 'Rear Upper Cranked Conductor Tube' in the maker's spare parts list and cost 3s. 3d. It fed the seed into the 'Rear Lower Coulter Tube'. Just below this second tin the top of the coulter can be seen.

LEFT: A Suffolk type drill which was a larger and later development of the Kent drill. It made use of the cup feed system. Among the many manufacturers of this type of drill were R. & J. Reeves, Westbury, Wilts.; F. Walker, Tithby, Notts.; W. Rainforth, Britannia Iron Works, Lincoln; W. Tasker, Waterloo Iron Works, Andover, Hants.; A. W. Gower, Market Drayton, Salop; and Thos. Holyoak, Narborough, Leicester.

HAND MACHINES

Until the nineteenth century there were many farm tasks which could be performed only by hand. Engineers did not confine their efforts to making work in the field easier and all kinds of small machines were designed to reduce the time needed to perform given tasks.

Some machines—such as the chaff cutters —had large exposed knives which could sever a misplaced hand or limb. In the early nineteenth century there were no regulations regarding safety and little attention was given to the hazards the operator had to encounter. In the 1840s even the smaller machines for cutting roots or crushing linseed, oats, barley or beans frequently had their working parts exposed. Later versions enclosed most of the moving parts and by the 1870s more thought was given to the safety of the user.

LEFT: *A mechanical wool clipper which replaced the ancient hand shears. Power was provided by a boy turning the handle. In 1900 sheep shearers were paid from 4s. to 5s. per score.*

BELOW: *Chaff cutter with two curved blades (one missing). Some of these early machines were dangerous to use.*

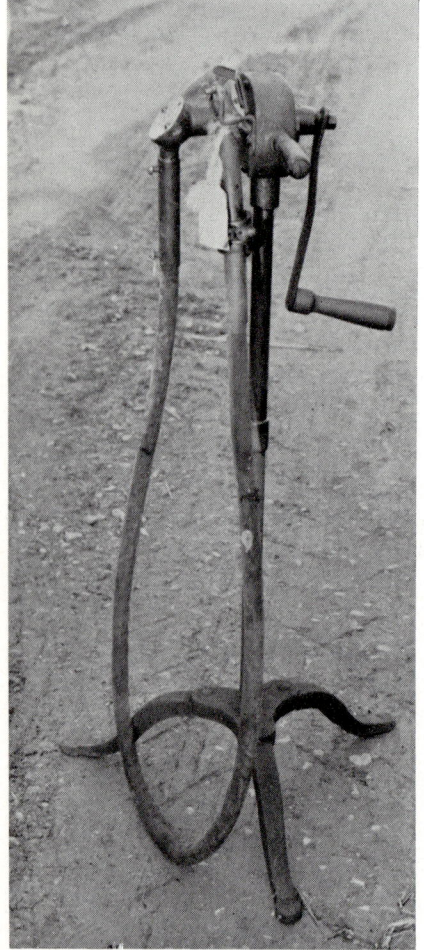

TOP RIGHT : *A root washing machine. The potatoes or other roots were placed in the cylinder which had its lower portion in a trough of water. As the cylinder was rotated the roots were washed and expelled in the manner shown.*

RIGHT : *Blackstone's (of Stamford) Improved Root Cutter. The roots used for animals had to be prepared. This machine, with its cutting parts enclosed, was arranged so that when the handle was turned in one direction it provided broad slices. By turning the other way it provided narrow slices which were more suitable for sheep.*

BOTTOM RIGHT : *A horse drawn double drill turnip and mangold sower with rollers of a characteristic shape. In 1894 such a machine cost £6—carriage paid! In the north of England machines of this kind were made by William Elder of Berwick-on-Tweed.*

BOTTOM LEFT : *A well worn grindstone.*

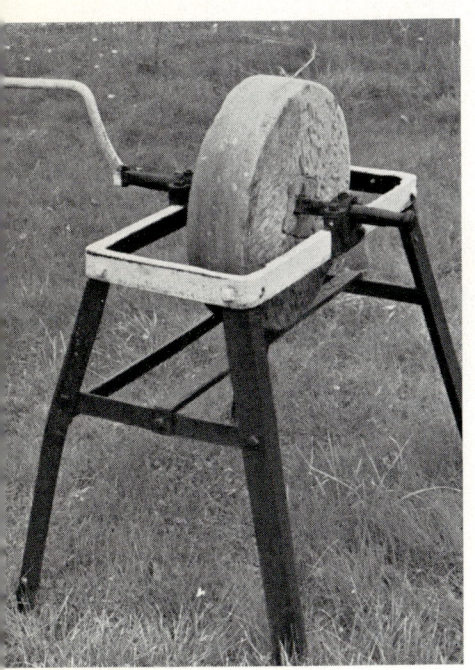

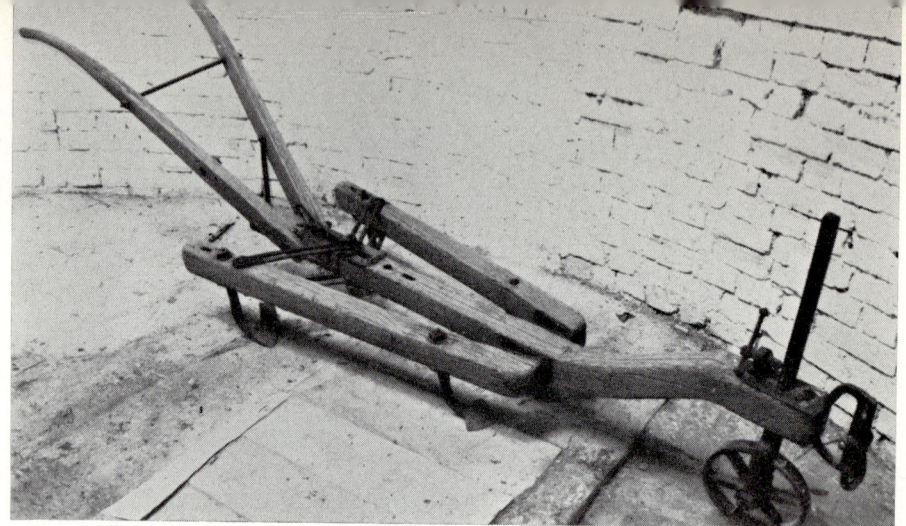

HOEING

With any growing crop, weeds also appear. In the early stages of germination some weeds will quickly smother the intended crop if left unchecked. The seed drill set out regular rows with fixed spaces in between. This allowed the use of another mechanical device which could uproot any weed growing in the spaces separating the rows. Once the weeds were exposed to the elements their chance of survival was greatly reduced. All kinds of ingenious hoes, scarifiers and scufflers were manufactured in the last century.

The quality of the crop depended to a great extent upon the quality of the hoeing it received. Horse drawn hoes did not completely replace the hand hoer: both methods had their place on the farm.

TOP : *An adjustable hoe from Sussex made with a minimum of iron work—suggesting an early origin in a rural workshop. Factory-made alternatives in iron followed the same design. (Norton's Farm Museum.)*
BOTTOM : *Another hoe variation in iron. Notice the spanner and spare hook on the handle.*

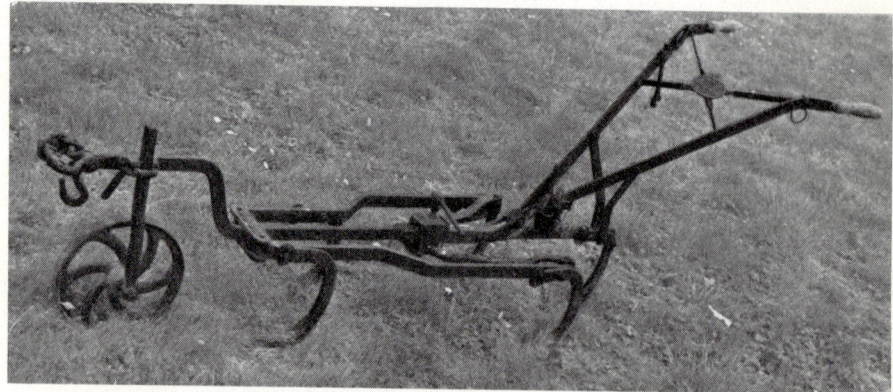

HAYMAKING

The gathering of the hay crop was an important part of the farmer's year. A good deal depended upon the quality and volume of the hay taken home to the haybarn. Enough had to be stored away to feed stock during the winter months and it was a happy farmer who had sufficient and some left over to sell. Although machines galore came to dominate haywork it is worth remembering that for well over a thousand years haymaking was the province of men, women and horses.

TOP : *The many stages involved in threshing rye-grass in the field : (a) the horse brings the haycocks; (b) a woman rakes up behind; (c) the hay is placed on an old door; (d-e) threshers at work with flails; (f-g) the threshed hay is thrown on a field gate and shaken out to allow the seed to drop below on to a canvas sheet; (h) threshed hay is thrown aside; (i) a new rick is made; (k) the hay seed is riddled by hand; (l) heap of hay seed ready for sacking; (m) seed in sacks; (n-o) completed ricks, called 'colls' in the north; (p) ladder; (r) spare rake; (s) refreshment for the ten workers — a reminder that haymaking is thirsty work.*
BOTTOM : *An 'Albion' mower at work. These well-known machines were made by Harrison, McGregor & Co., Leigh, Lancs.*

TOP : *This hay-sledge was made by John Wallace and Son, Glasgow, in the 1880s. The flat platform tipped backwards so that the edge of the haycock could be lifted on to it. A chain was wrapped around the haycock and as the operator pulled on the rope wound round the windlass the burden was slowly pulled on board. As the hay moved across, the platform fell back into the position shown in the illustration.*

CENTRE : *A wooden hay-sweep designed to be pulled by a horse.*

BOTTOM : *After the hay had been cut it was turned with the horse drawn 'hay making' machine or hay tedder. Some tedders had a protective guard to prevent injury if the driver fell from his seat. The spindle on which the revolving tines are mounted was driven by gears mounted on each land wheel.*

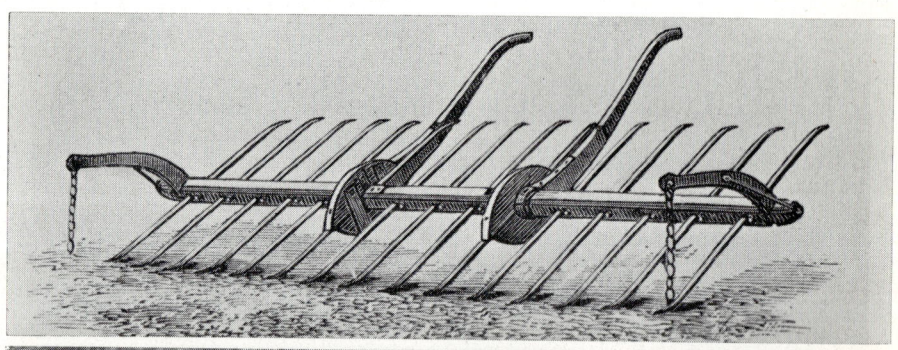

HARVEST

The climax of the farmer's year came with the harvest, when all available hands were sent into the fields. From the farm worker's point of view the extra money the family could earn at harvest time was a matter of great importance. When the annual battle with time, daylight and the weather had been won and the last sheaves carried home, then came the traditional harvest supper. Towards the end of the last century great advances were made by engineers and the ordinary mower was transformed into a machine of great complexity. Since then the trend has continued and the massive combine harvesters we see in today's harvest fields are the inheritors of the intricate technology created by countless, often anonymous, Victorian engineers.

ABOVE : *A self-raking Hornsby reaper. It delivered cut corn in bundles ready for the hand binders but clear of the track it had followed. The way was therefore clear on its next circuit of the field and it did not have to wait for the hand binders to catch up.*

LEFT : *This front view of a restored reaper, shows the cutting bar and the way in which the sails swept the cut corn clear of its track.*

TOP : *A reaper at work near Mordiford, Hereford about 1930. (University of Reading Museum of English Rural Life.)*
BELOW : *In the 1880s the binder made its appearance. It was a considerable advance on the reaper which left the cut corn to be bound by hand. The binder, with its paddles (sails) to push the corn towards the cutting bar, took each swathe into its charge, bound the sheaves and discharged them ready for stooking. It added speed to the operation. With a binder a farmer could cut the corn and get it out of danger more quickly if bad weather threatened.*

ABOVE : *A baling press of vast dimensions. This machine could be used for hay or straw. Bales required less space to store than loose material.*

BELOW : *An old threshing box at work again. Unthrashed corn is fed into the top of the machine. It passes round a drum which removes most of the grain from the husks. The straw is expelled from the rear of the machine and an elevator carries it to the stack. Inside the machine the grain falls through a series of riddles. A fan helps to separate the chaff before the cleaned grain is carried, via an elevator, to the sacks.*

ABOVE : *Once the corn had been gathered it was ready to be ricked. This elevator was designed to be worked by a horse attached to the horse gear (right).*
BELOW : *Steam engine, threshing box and elevator complete a scene which was once common in the countryside, but is now a novelty. The plastic sacks of coal and the oil drum date the picture, which was taken in 1972 at Marsh Gibbon, Buckinghamshire.*